Falcon and Owl

A Bird Book for Kids™

By Novare Lawrence

Nada Bindu Publishing Co.

The contents of this book previously appeared in the digital-only editions *Falcon: A Bird Book for Kids* and *Owl: A Bird Book for Kids*.

First Print Edition – March 2015

ISBN-10: 1633070093
ISBN-13: 978-1-63307-009-7

Published by:
Nada Bindu Publishing Co.
Carson City, NV 89703
Website: www.nadabindupublishing.com
Email: inquiries@nadabindupublishing.com

CONTENTS

Falcon   1

Owl   27

Falcon

The Peregrine falcon is the fastest bird in the world. In fact, peregrines are the fastest living thing on land, in the sea or in the air. They can dive through the air at speeds of 200 miles per hour (322 kilometers per hour). The fastest speed ever measured for a peregrine falcon was 242 miles per hour (389 kph). That is very fast for a bird that only weighs three pounds (1.4 kilograms) as an adult.

The fastest bird species: the Peregrine Falcon

To be fair, falcons only reach such top speeds by flying up very high, then diving downward using gravity to help them build up speed. When they fly that fast through the air, special eyelids protect their eyes and special bones in their nose help them to breathe. In normal level flight, peregrine falcons only fly around 60 miles per hour (97 kph) using just their wings alone. That is still as fast as a cheetah can run. The White-throated Needletail, or Swift, however, is the fastest flyer using wings alone and can fly over 100 miles per hour (161 kph) in level flight.

In flight: a powerful falcon versus a streamlined swift

Peregrine falcons are one of 37 types of falcons found around the world, which include the Kestrel, Gyrfalcon, and Hobby families. Within the peregrine falcon family itself, there are about 19 different varieties of peregrines. Peregrine falcons range from one to two feet (30 to 61 centimeters) in body length with a wingspan from 2-1/2 to 4 feet (76 to 122 cm). The tail feathers can extend another 5 to 7-1/2 inches (13 to 19 cm). Female peregrines are up to 30% larger than the males.

Peregrine relatives: the Gyrfalcon, Hobby and Kestrel

The peregrine falcon is a very distinctive and attractive bird with bluish-grey feathers on its head, back and outer wing surfaces. Its chest is white or tan with a darker grey or black dashed pattern that merges into lines as it flows down into its tail feathers and under-wing feathers. The beak often has a splash of yellow around the nostrils which match the color of its feet.

The distinctive colors and patterns of a Peregrine Falcon

Falcons are part of a family of birds called raptors, meaning birds of prey. Raptors are hunters relying primarily on a diet of meat as opposed to plants. While raptors in general may eat fish, small reptiles, rodents or other small mammals, peregrine falcons most often eat other birds.

Other raptors include Owls, Hawks and Eagles

Raptors can be distinguished from other birds by three things. First, they have excellent eyesight which helps them to find prey. While flying, you can survey a lot of territory without using up too much energy but you have to be able to see what is beneath you in the air, on the ground or in the trees as you glide overhead. A falcon's eyesight allows it to do this. It can see at least 2-1/2 times better than a human. Falcons will also sit on the tops of trees, on telephone poles, on rocky cliffs or on the sides of buildings and scan their surroundings for prey.

A Peregrine has excellent eyesight

The second thing that distinguishes raptors is their talons. Raptors have sharp, strong talons to help them catch and hold their prey. Many raptors catch prey as they fly or swoop down out of their hunting dive. They may catch a fish in the water, a rodent in a field, or a bird out of the air, and carry it away to eat where they won't be bothered by other animals. Raptors with young chicks will take captured prey directly to the nest.

A Peregrine has strong sharp talons

The third and final thing that distinguishes raptors is their sharp, hooked beak. Talons are used mostly to capture and hold prey. The beak is used to quickly kill the prey, usually by severing its spine. The beak is also effective at removing fur, scales or feathers depending on the type of animal that is caught. These three tools -- keen eyesight, strong talons and a hooked beak -- are how raptors hunt to feed themselves and their chicks.

This Peregrine shows its specialized raptor beak

In addition to falcons, other raptors include eagles, hawks, owls, kites, harriers, vultures, buzzards and ospreys. Most of these are hunting birds searching for live prey but vultures in particular only look for dead or dying animals. When food is scarce however, even hunting raptors may try to take advantage of prey killed by another bird or animal.

Even vultures display the three traits of a raptor: keen eyesight, strong talons, and hooked beak

Male peregrines are smaller than the females. Both are effective hunters with speed, strength and all the raptor characteristics working for them. As one of the more powerful falcons, the peregrine hunts mostly small and medium sized birds or bats but may include small mammals and reptiles if birds are scarce. Peregrines will often grab other birds right out of the air which shows just how capable and fast they are in the sky.

A Peregrine after a successful hunt

On average, peregrines live to be around 15 years old but have been known to live to age 20. Their primary threat is from eagles and larger owls which may hunt them. Back in the 1970s, falcons were declared an endangered species because of the widespread use of the pesticide DDT which harmed their ability to lay eggs. Since DDT was banned, restoration projects have protected falcon nesting areas and supported repopulation efforts. Today, the number of falcons in the wild has mostly recovered with this assistance but it is a reminder that what humans do can greatly affect other animals.

A healthy Peregrine in the wild

When breeding season comes, falcon courtship displays can be quite acrobatic. Both the male and female will spiral and dive around each other as they fly together, showing off their skills. The male sometimes passes prey it has caught to the female, in mid-air. The male holds it in his talons while the female flies upside down beneath him to grab it with her talons. Other raptors may also pass prey in mid-air as a game or as training for younger raptors.

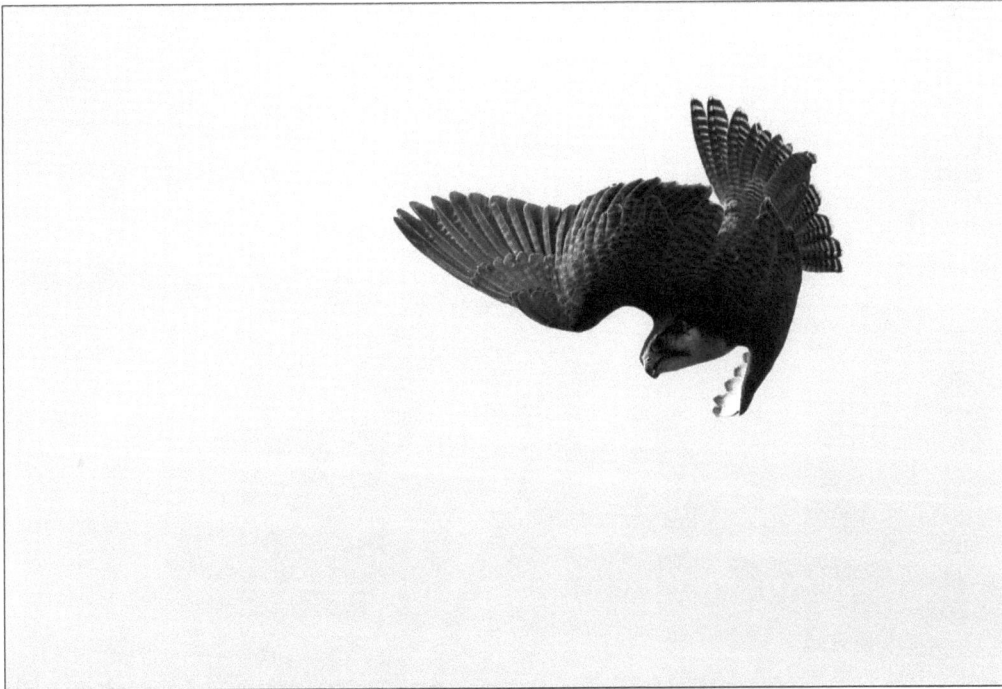

A Peregrine turning almost upside down in mid-air

Male and female peregrines mate for life. They almost always return to the same area to nest each year and may even use the same spot on which to build or restore their nest. Falcons like to build their nest up high because it is easier to protect their young in such a location. Falcons may nest in the hollow of a tall tree or on a cliff or rocky shelf accessible only from the air.

A Falcon nesting in a bird-sized cave on a rocky cliff

Falcons may even build a nest on the window ledge of a tall building or on the girders of a bridge. Peregrine falcons are one raptor that has made the adjustment to city life, especially where pigeons are present. In areas like the Arctic, which doesn't have many ground animals, peregrines may build their nests on the ground. Even there, they will try to nest on a mound of dirt or rocks which allows them a better lookout point to watch for any approaching threats.

A Peregrine in the city

The female falcon lays from 1 to 5 eggs which hatch after about a month. The male may sit on the eggs while the female hunts but the female peregrine does most of the egg incubation. Both parents will hunt for food for the chicks after they hatch and defend the nest together. When the chicks are very small, one parent often stays on or near the nest while the other is away hunting. Parents may fly up to 15 miles (24 kilometers) away from the nest when hunting to feed the growing chicks.

A Peregrine sitting on her eggs in a ground nest

Peregrine falcon chicks are covered in very soft white downy feathers to keep them protected as the parents leave them to hunt. The parents will catch prey and bring it to the nest, tearing off small pieces of meat to feed the chicks. When the chicks are strong enough, they will begin to do this for themselves, having learned by watching their parents. Any bones, feathers or other indigestible parts that are swallowed are regurgitated later as a small dry ball.

A Prairie Falcon chick

Falcon chicks fledge around 45 days after hatching. Fledging means that the chick has developed all of its adult flying feathers and is ready to leave the nest on its own. Most of these fledglings still need to build up their flying and hunting skills and do so by following their parents around for a while. The parents may feed them at first but after a month or two they will finally push their offspring away, forcing them to be independent.

A Peregrine going off on her own

Many birds, including falcons, use a process called imprinting. Imprinting is where a newly hatched chick bonds with the first living thing it sees. Many parent birds incubating their eggs will coo or chirp to the eggs. As it matures prior to hatching, the developing chick will chirp back from inside the egg. Then, as it hatches, the parent will chirp again to catch the attention of the chick and complete the imprinting bond as the chick sees the parent for the first time. After that, the chick will follow and learn from its parents only. This sound connection is especially important for birds that nest in colonies so that parents and chicks can find each other when the parents come back after hunting.

A Peregrine parent on the ground near its nest

People who practice falconry use this imprinting process to form their own bond with young falcons or other raptor chicks. Falconry is a long practiced discipline that originated in Asia over three thousand years ago but has been adopted throughout history into many cultures. It is still popular today. A falconer imprints with a hatchling and raises it. As it gets older, the falcon will be taught how to hunt prey but also how to return it back to the trainer. In the past, this was how falconers fed their family. Today, it has also become a sport that tests the training and skill of both the human trainer and the falcon's abilities against other competitors.

A Falcon with her human trainer

Falconry requires a lot of dedication on the part of the trainer as they become the replacement parent to the young falcon. In most countries, you must have professional training and obtain a license in order to become a "falconer." This is designed to protect both you and the falcon from harm. Falcons, hawks, eagles and owls can be used in falconry but traditionally it is falcons and hawks that most falconers train.

Falcon and trainer in full protective gear

There is protective gear that both the trainer and the falcon wear. The trainer protects his or her arm with a thick leather sleeve and glove. The bird is trained to take off from and land back onto that protected arm. Before a hunt or when being moved, the bird wears a protective hood that helps to keep it calm, especially in noisy environments and around other birds and people. The bird also wears leather straps on both feet that allow the bird to be tethered to a perch so that it cannot fly away. Each bird may have a small bell, or a pair of them, attached to its ankles to help the falconer track the bird when it's flying. Birds are always tagged for identification and may have a temporary GPS transmitter attached when out hunting with the falconers.

A Falcon with its hood in place

In falconry, falcons and other raptors do not fly away from their trainers when released to hunt. Through the imprinting and training process, the birds learn that they will be well fed and cared for by their trainers so they always return with their catch. Even though the birds become very used to their trainers, they never become tame like a house pet. You should never approach any trained bird without the trainer being present and giving permission. The best place to see a falcon or other raptor up close is at a zoo or an animal park. They will often have a show where you can see raptors flying as directed by their trainers.

A large Harris hawk giving a demonstration at a fair with leather straps and bells on

Falcons and other raptors are the top hunting birds in nature yet they work well as partners with people in hunting and contests of skill. They are intelligent and very capable and have even been used by trainers to chase other birds away from airports and airplanes. Training and exercise is a daily activity as the birds depend completely on their dedicated trainers.

A Peregrine in training

Although it is smaller than many raptors, the peregrine falcon is the best example of strength, skill, intelligence and precision as it dives through the air faster than any other living thing.

Owl

Owls are one of the most easily recognized birds despite the fact that there are over 200 types of owls. This is because owls share a common set of features: a blocky body shape, a flattened face and large, forward looking eyes. Most other birds have a sleek streamlined shape from head to tail, their faces are pointed and their eyes are positioned on the sides of their heads.

A Great Horned Owl compared to a Kestrel Falcon

Owls are generally stocky, with no visible neck. When perched on a branch, they look like a big feathery box. Even though you can't see its neck, an owl can turn its head a full 270 degrees to look over its own shoulder and see what's behind it without moving its body. Most people can only turn their heads about 180 degrees, from shoulder to shoulder.

An Ural Owl looking over its shoulder

Owl necks are also flexible from side to side. Sometimes you'll see an owl tilt its head far to one side, then the other, never moving its eyes away from whatever it's looking at. This helps the owl to see an object better, figure out what it is and judge just how far away it is.

An owl's neck is very flexible

An owl's eyes are very big in relation to its head. With such big eyes, owls can see very well at night, which is when most owls hunt. Owls have a flattened face, like us, so their eyes face forward but an owl's eyes don't move. They can't roll their eyes from side to side the way that we can. To see something, an owl has to turn its head to face the same direction as the object that interests it. Once they focus on something, their eyes are like binoculars that allow them to see small animals some distance away.

A Barking Owl's big eyes

Owls have three eyelids in each eye. The first eyelid is for blinking which keeps the eye moist. The second eyelid cleans the eye to keep it healthy and free from dust or other things which the owl may encounter during flying. The third eyelid is used when sleeping. Although owls can see distant objects well, they cannot focus on close-up objects. They often use the small feathers around their beak to feel or touch things on the ground.

Owls use the small feathers around the beak and eyes to feel objects up close

Owl ears are also specialized compared to other birds. The tufts of feathers above the ears that some owls have are not really ears. They are a group of feathers that move up and down and usually express emotions such as alertness or being tired. An owl's real ears are positioned on the side of their head. The left and the right ears are different sizes and one of the ears is positioned slightly higher up on the head than the other. This helps an owl to pinpoint where a sound is coming from.

Although a Tawny Owl has no ear tufts and an Eagle Owl does, both have excellent hearing

The other thing that allows an owl to hear so well is the shape of their face. An owl's flattened face helps to capture sound for its ears, like a radar dish, which makes its hearing about ten times better than ours. Owls often rely on their hearing to detect prey first before turning their heads to see it with their keen vision. With their big eyes and superior hearing, owls can hunt very effectively at night.

You can see the round dish-like shape of this Oriental Scops Owl

Owls belong to the family of birds called Raptors, just like falcons, hawks, eagles and vultures. Raptors have three distinguishing features among birds -- excellent eyesight, sharp talons on their feet, and a hooked beak -- and live on a diet of meat instead of plants. Owls have special feet compared to other raptors; they have two toes facing forward and two toes facing backward. This gives owls a very strong grip for capturing prey. Most other raptors have a three and one arrangement of toes. Depending on an owl's size, it may eat rodents and other small mammals, reptiles, fish, birds and insects.

Two front-facing toes for Owls, three front-facing toes for Eagles

Owls have two additional things going for them that enhance their raptor hunting capabilities. Unlike parrots or peacocks, owls have very dull-colored feathers which help the owls blend into their surroundings, especially in the darkness of night. Even during the day, it can be hard to see an owl in its natural habitat.

Owls like these Tawny, Northern Saw Whet and Scops Owls can be hard to spot during the day

The wing feathers of an owl are the second important thing that helps it while hunting. Right along the leading edge of an owl's wings are special feathers that muffle the sound the wings make as they push through the air during flying. Owls also have a softer layer of feathers, like velvet, that coats the outside surface of their wings. It acts like soundproofing to reduce the sound of the wings moving through the air. When hunting, owls are very hard to see and very hard to hear.

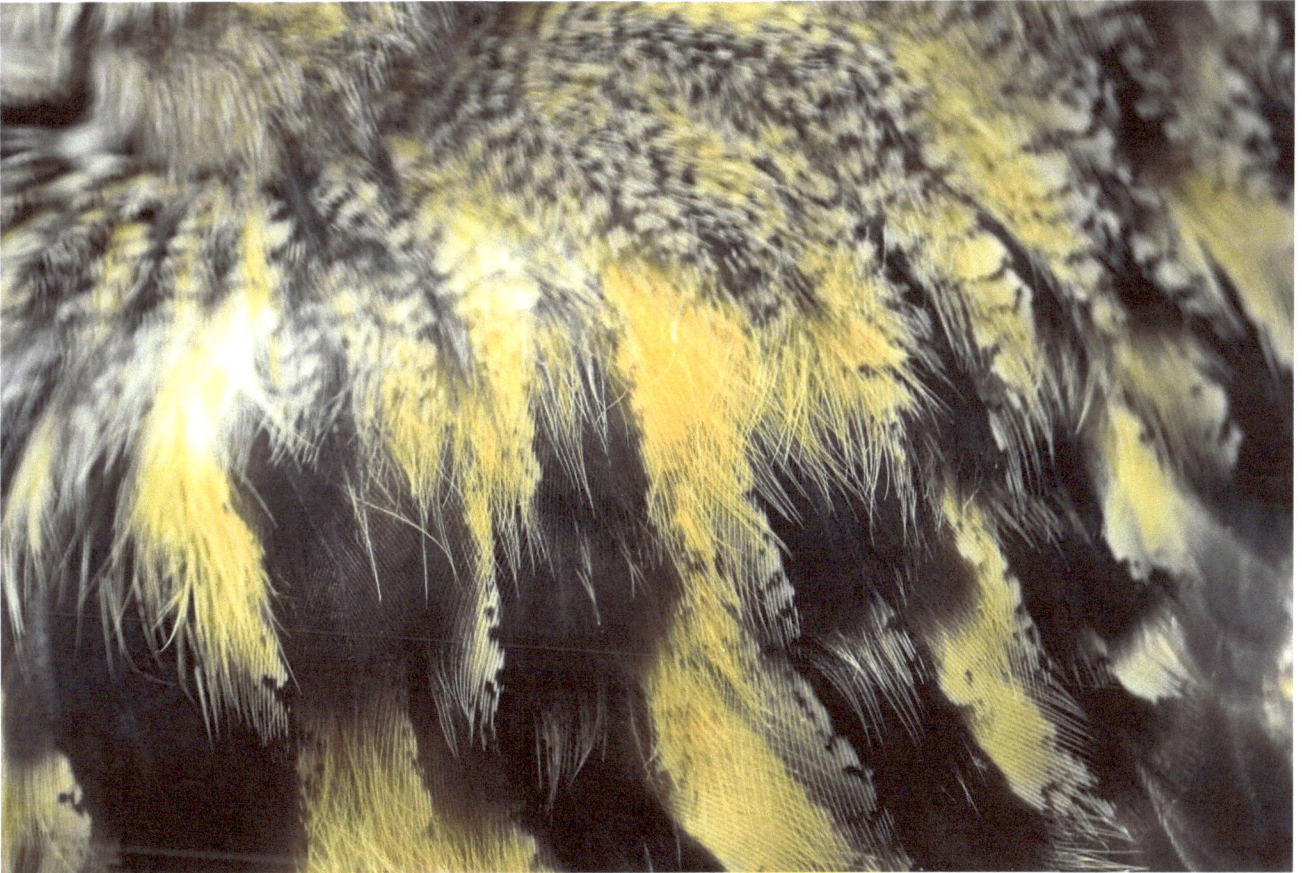

The soft coating on an Eagle Owl's wing feathers

Owls have no teeth so they use their sharp beaks to tear off smaller pieces of meat from their prey or, if they can, they simply swallow it in one gulp. Their stomach is very specialized and separates meat from feathers, bones, fur or other things that cannot be digested and compresses those into a small pellet-sized lump. The owl then regurgitates the hard, dry pellet through their mouth the way a cat might cough up a hairball. You can usually find these owl pellets around their nesting areas.

A dry owl pellet of feathers and bones

Owls belong to a group of birds that includes about 205 different species divided into two basic groups, barn owls and true owls. There are 16 species of barn owls and they are all medium-sized for an owl. True owls, with about 189 different species, vary widely in size.

A Barn Owl and two true owls – the big Great Grey Owl and the small Tawny Owl

Among the True Owls, there is a great range in size. The Elf Owl and Pygmy Owl are the smallest of the owls. An adult may weigh only 1 oz. (31 grams) and sit only 5 inches (13.5 cm) tall. The biggest owls are the Great Grey Owl, Eurasian Eagle-Owl and the Fish Owl. These owls are all in the range of 10 lbs. (4.5 kg) in weight, 32 inches (82 cm) from head to tail, and big wings that span up to 6.5 ft. (2 m) from wingtip to wingtip.

A big size difference: the Pygmy Owl vs. the Eagle-Owl

For their size, owls don't weigh as much as you would expect. The reason they look so big is because of their full feathers which hide their neck and even cover their legs. Owls don't fly that fast either since they don't rely on speed when hunting, they rely on stealth. Owls tend to hunt from a perch, using their vision and hearing to locate prey, then they silently swoop down to catch it with their strong talons. Even a medium sized owl's grip is stronger than that of a person's hand.

An owl looks big but it has a lot of fluffy feathers

Most owls are nocturnal which means that they are awake at night and sleep during the day. This is one reason why most people have never seen an owl even if they live near farming or wilderness areas. Hunting at night also means less competition as other raptors, such as hawks and falcons, will only hunt during daylight hours.

Barn Owls at night, hunting to feed their owlets

There are a few species of owls that are not nocturnal. One of them is the burrowing owl which is a small owl from 7.5 to 11 inches (19 to 28 cm) long. Burrowing owls still do most of their hunting at night but are more active during daylight hours than most other owls. Burrowing owls live in burrows underground. The burrows may have been borrowed from other animals, like prairie dogs, or the owls may dig them on their own. Burrowing Owls spend so much time on the ground that their legs are longer than most owls and free of feathers. They are quite good at running short distances. If a predator approaches them, they can easily run back to their burrows.

A Burrowing Owl watches over her owlet

As with many other bird species, female owls tend to be bigger than males. Female owls lay three to four eggs on average but some species have been known to lay up to a dozen. The female lays her eggs over several days and the eggs hatch one by one in the same order after about a month of incubation. Baby owls are called owlets and the first one hatched usually gets the best food because it is older and bigger than its brothers and sisters that hatch later.

A Great Grey Owl nest with eggs and very young owlets

Different owl species can have very different nesting styles although owls don't build nests piece-by-piece the way many other birds do. Some owls look for rocky ledges with small crevices or caves where they can safely lay eggs. Some use nests in trees made by other birds or animals. Others may look for hollowed out tree trunks or stumps. And some dig burrows in the ground. Owls will also take advantage of space in barns and attics as well as in specially made owl boxes - as long as other predators cannot easily get to their nests.

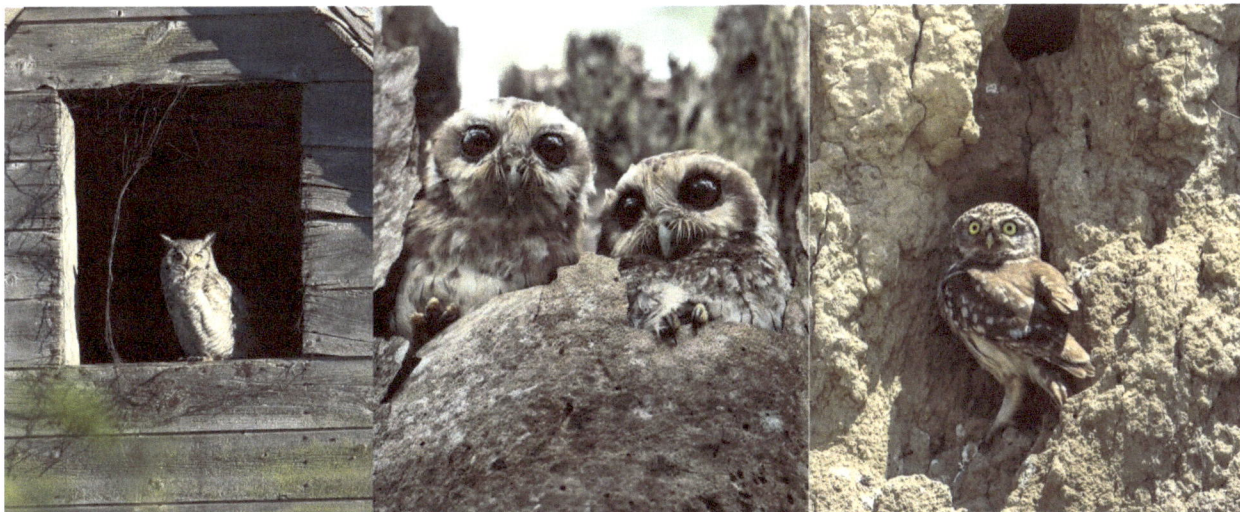

A Great Horned Owl in a barn, two Bare Legged-Owls in a tree trunk, a Burrowing Owl in a hillside

During breeding season, owl parents hunt almost continuously, not just at night, in order to feed their hungry owlets. This is good for people. Owls keep rodent populations low in farming areas and help in many rural and even suburban areas to control small animals that may harm plants and homes. If you know that there are owls in your area, you might even be able to attract owls to your yard. If you can build and place an owl box in your yard, far away enough from your house or anything that might disturb the owls, they might select it as a nesting place.

A Great Grey Owl sharing a successful hunt with one of its hungry owlets

In order to plan your owl box, you need to find out what types of owls you have in your area to determine the correct size. It is easy to check on the internet for this information. You need to be sure that you have a place to put the box. It will need to be away from any regular activity or noise from your home or other nearby yards and be mounted at least 10-12 feet off the ground. An owl box may be placed in a tree, mounted on top of a pole or mounted to the side of a structure like a barn.

A fancy owl box mounted in a tree for a Barn Owl family

Owl boxes can be very plain, just four walls, a roof and a floor that has some small drain holes around the bottom corners in case rain gets in. The entrance hole must be large enough for the parents to enter, and there is usually a perch below the hole that they can grab onto when entering or exiting. The hole should be positioned in the top third of the box so that eggs and young owlets are well hidden in the bottom portion of the box. An adult with the proper tools should supervise and help with the building and mounting of your owl box.

A simple bird box mounted in a birch tree.

Only a few species of owls migrate regularly like other birds. Some owls may follow prey that relocate during the different seasons of the year but most stay year-round within the same general territory. Owls often return to the same nesting locations each year. Once a mated pair finds an owl box, they will most likely return to it each year to lay eggs and raise their young. Owl species are found all over the world with the exception of Antarctica and some remote islands.

Owls thrive in all different climates

Owls have been familiar to people and used not just in falconry but also as symbols in human cultures for thousands of years. In Africa, Asia, Greece, Europe and the Americas, owls have represented different things to different peoples. Owls are found in old paintings and wall carvings, on totem poles and painted pottery and even on ancient coins.

Owls in ancient Egyptian carvings, on ancient Greek coins and Native American totems

Owls are said to represent good qualities like wisdom, which comes from ancient Greece over 2500 years ago. Perhaps the big eyes and steady glare of owls convinced some people that owls must be very intelligent. Owls certainly are curious and are quick to lock their stare onto anything that interests them. Whether they are truly wise or not, we can simply appreciate owls for how unique, beautiful and capable they are.

A curious Spectacled Owl

Owls are smart and have long been used in the traditional art of falconry, along with hawks, eagles and all species of falcons. It is not as common to use owls because most owl species are night hunters but you can find trained owls in zoos and animal shows. Trained bird handlers wear the same protective gear as falconers to protect themselves and their birds from harm. The protective hoods that falcons wear are not used with owls due to the shape of owl heads.

The falconer wears a protective glove. The owl wears straps on its legs

Owls are capable of making many more sounds than the "hoo hoo" hoot sound we usually think of. Their range of calls include screeches, whistles, barks and hisses, some of which can be heard up to a mile (2.2 km) away. Their cry can sound like a baby crying or a dog barking. Some of the smaller owls that are often found in petting zoos can almost purr like a cat when you stroke their feathers.

What sound is this Eagle Owl making?

Owls are also very enjoyable to look at because their eyes and head movements can be so expressive. There is such a variety of shapes, sizes and expressions to see. Despite their differences, however, you can always tell by their shape and common features that it is an owl.

Two owlets and a Long-Eared Owl

So if you ever visit a forest or any area in the country with lots of trees, you might want to watch out for owls and listen for their distinctive "hoo hoo". Even though they prefer the nighttime, you might just see one if you look very closely.

Peek-a-Boo!

ABOUT THE AUTHOR

Novare Lawrence loves researching and writing books about Nature. She shares the knowledge and beauty of our natural world with kids young and old hoping that we will all do our part to help preserve our planet and all the wonderful species upon it.

You may learn more about her books and the *A Bird Book for Kids*™ series at her website:
ABirdBookforKids.com
And at
NadaBinduPublishing.com

A Bird Book for Kids™ Books
by Novare Lawrence

Digital:
- Hummingbird
- Ostrich
- Falcon
- Owl
- Condor
- Peacock
- Quail
- Pelican

Print:
- Hummingbird and Ostrich

www.ingramcontent.com/pod-product-compliance
Lightning Source LLC
Chambersburg PA
CBHW060816270326
41930CB00002B/62